Panzerjäger

Improvisations, Combinations on Captured Chassis, Marder I and II, Prototypes, etc.

by Horst Scheibert

Left: A Czech 47 mm Pak on the chassis of a French Renault 35 R tank. In 1942-43, 184 of these were built. They saw service in Russia and France.

Schiffer Military/Aviation History
Atglen, PA

Sources:

Federal Archives, Koblenz (BA)
Library for History of the Times
Scheibert Archives
Podzun Archives
Hans-Joachim Schröder Archives (Sch)
Hans Ritter Klippert

Translated from the German by Ed Force

Front cover artwork by Steve Ferguson

Printed in China.
ISBN: 0-7643-0395-3

This book was originally published under the title, *Panzerjäger-Improvisationen, Zusammenbauten auf Beutefahrgestellen, Marder I und II, Prototypen u. a.* by Podzun-Pallas Verlag GmbH.

We are interested in hearing from authors with book ideas on related topics.

Published by Schiffer Publishing Ltd.
4880 Lower Valley Road
Atglen, PA 19310
Phone: (610) 593-1777
FAX: (610) 593-2002
E-mail: Schifferbk@aol.com
Please write for a free catalog.
This book may be purchased from the publisher.
Please include $3.95 postage.
Try your bookstore first.

A Kfz 69 (Krupp L 2 H 143) with a 37 mm Pak mounted on it. A troop improvisation; not even the wheels have been removed.

Panzerjäger

No service arm of the Wehrmacht changed so quickly and thoroughly during the war as the anti-tank forces. Out of the necessity of facing enemy tank masses and strength, the 37 mm Pak gun towed by a truck, the only such gun available at the start, was developed into numerous new weapons and systems that culminated in the heaviest German armored vehicles that saw action on the battlefield, the 76-ton Jagdtiger. Necessity did not beget such extensive improvisations for any other weapon—with the possible exception of the Flak guns—and weapons originally built for other purposes—such as the assault artillery and anti-aircraft forces—were put into action as anti-tank guns.

Above: A light armored scout car (Sd.Kfz. 221) with built-in 28 mm heavy Panzerbüchse 41. This vehicle reached the front in 1941; in 1939 there were a few vehicles with the 28 mm heavy Panzerbüchse beside a machine gun.

Right: A light Schützenpanzerwagen (Sd.Kfz. 250/11) carrying a heavy Panzerbüchse 41 (28 mm). It was used by the SS Panzer Division LAH in southern Russia in 1942.

28 mm Heavy Panzerbüchse to 50 mm Pak Inprovisations and Factory-built Vehicles

In the Polish campaign it had already become clear that in mobile warfare an anti-tank gun towed by a limber or a tractor often could not be made ready to fire quickly enough. As a result, the troops, improvising at first, mounted this weapon, and then others suitable for anti-tank use, on motor vehicles or chassis already available—whether German-made or captured. The industry did the same later. In terms of time, the well-known 37 mm Pak gun appeared first, then the Czech 47 mm gun taken over in 1939, and in 1940 the captured French 47 mm Pak, as well as, finally, the 28 mm Heavy Panzerbüchse 41 and the 50 mm Pak.

Besides numerous improvisations made by the troops, the industrial productions—listed by caliber—were as follows:

28 mm	s. Pz.-Büchse 41 on light armored scout car, Sd.Kfz. 221.
28 mm	s. Pz.-Büchse 41 on light armored scout car, Sd.Kfz. 222.
28 mm	s. Pz.-Büchse 41 on light Schätzenpanzerwagen. Sd.Kfz. 250/11.
37 mm	Pak L/45 on one-ton light towing tractor.
37 mm	Pak L/45 on light Schützenpanzerwagen, Sd.Kfz. 250/10.
37 mm	Pak L/45 on medium Schützenpanzerwagen, Sd.Kfz. 251/10.
37 mm	Pak L/45 on Infantry Tractor UE (f).
37 mm	Pak L/45 on Bren Carrier (e) chassis.
37 mm	Pak L/45 on Artillery Tractor (r) STZ-Komsomoletz.
47 mm	Pak (t) on Panzerkampfwagen I B chassis, Sd.Kfz. 101.
47 mm	Pak (t) on Panzerkampfwagen 35 R (f) chassis.
47 mm	Pak 181 or 183 (both f) on armored Lorraine Tractor (f).
50 mm	Pak 38 L/60 on Borgward VK 301/302 ammunition tractor chassis.

This Sd.Kfz. 250/11 has an MG 34 in anti-aircraft position. This version reached the troops in 1943. The crew consisted of three men.

Upper and lower right: Both photos show the same limber truck (Kfz. 69) with a 37 mm Pak set on it. It is recognizable as an improvisation by the wooden blocks between the spare wheel rack and the rear shields, as well as the pillows padding the folded-down windshield. Surely the driver had to get out during firing, and yet it was all faster, nimbler and higher than a towed Pak gun.

Above: A troop improvisation: a Russian Panzerbüchse on a medium Schützen panzerwagen (Sd.Kfz. 251). This gun, like the German Panzerbüchse, was only capable of penetrating lightly armored vehicles.

Upper left: A 37 mm Pak rigidly mounted on a British Bren carrier. There were very few of these.

Above: A 37 mm Pak on a French UE infantry tractor.

Left: The I./Panzerjäger 16 (16th Panzer Division) built this in the summer of 1941, mounting a 37 mm Pak with reinforced shield on a one-ton towing tractor, with a primitive armor plate at the front end. Here too, the folded-down windshield had to be padded to prevent shattering in combat.

Above: Another troop improvisation that combined the 37 mm Pak gun and the one-ton towing tractor.

Upper and lower right: These factory-built combinations mounted the 37 mm Pak on the light SPW 9Sd.Kfz. 250/10) and medium SPW (Sd.Kfz. 251/10). Each was made both with and without a shield. Here the 37 mm Pak is seen on both, one with and one without a shortened shield. Above is Sd.Kfz. 250/10, below 251/10. Usually the platoon leader rode in these vehicles.

These three pictures show the medium Schützenpanzerwagen Sd.Kfz. 251, Type 10, with a 37 mm Pak mounted on it. Unlike the picture on the previous page, showing the shorter shield or none at all, the vehicle shown here is the earlier type. The shield turned out to be too great a hindrance in combat field observation.

On this page are more action photos of the medium Schützenpanzerwagen with the 37 mm Pak, this time with the new short shield.

The right photo is of particular interest, as it shows the later version of the medium SPW with angled rear end.

A Czech 47 mm Pak on the chassis of the Panzer I B tank. This vehicle was designated Sd.Kfz. 101, and 350 of them were built. (Sch.)

The two upper pictures show the space for the crew, which consisted of three men. The upper right picture is interesting, as it shows the Panzerjäger I, as the combination of the Czech 47 mm Pak with the Panzer I was known, alongside a medium SPW with the older 37 mm gun.

This first Panzerjäger on a tank chassis was used by the Panzerjäger units of the Panzer divisions in 1940, and later (as of 1941) also by the infantry divisions. It could be found on almost all fronts from 1940 to 1942.

Because of its good gun—if one overlooks the small caliber—it was a useful tank destroyer in the first war years in France and Africa.
Opposite page:
Above: In Africa, at right in the winter, and at lower left in the summer in Russia. The crew usually wore (except in Russia) the black Panzer troop uniform—one reason being that all captured materials could be dyed black.

A Czech 47 mm Pak on the chassis of the French Renault 35 R tank. Another photo can be seen on the inside of the fron: cover. (Sch.)

These three photos also show the Panzerjäger on the Renault 35 R chassis. The two upper pictures were taken at a French training camp. Right: On the way to the invasion front in 1944.

Its weight was over ten tons and its crew consisted of three men. Its four-cylinder, 5.9-liter engine produced 80 horsepower. Thus, its power-to-weight ratio was around 9 and it was slow and relatively ponderous. (all BA)

Above: This French Lorraine tractor with a French 47 mm Pak (181) mounted on it offered the gun crew little protection. It was only a stopgap, thus there were few examples of it.
Upper right:
This 50 mm Pak 38 on a light towing tractor was surely a troop improvisation. In combat the driver had to get out, and the crew had no protection. Only a few of these are known to have existed.
Right:
The prototype of a 50 mm Panzerjäger consisting of a 50 mm Pak 38 L/60 mounted on the chassis of the Borgward VK 301/302 ammunition tractor.

Opposite page: This good photo shows design details of the Panzerjäger, which had a rather unfortunate shape and a fighting compartment open at the top, though it offered the crew plenty of space and protection. (Sch.)

This is a front view of the planned 50 mm Panzerjäger, of which large numbers were supposed to be built. The inferiority of the 50 mm gun compared to the 7.62 cm guns of the Russian tanks caused the plans to be cancelled.

Its weight was four tons, its crew numbered three men.
Below: A look into the fighting compartment.

75 mm Pak 40/1 on French Chassis

With the appearance of numerous Russian tanks, particularly the T-34, many combinations were made (see next section), including that of the German 75 mm Pak 40 L/48 with captured and newly produced French chassis. The type built most often used the chassis of the Lorraine tractor and was called "Marder I".

There were four types in all:

75 mm Pak 40/1 L/48 on Lorraine Tractor, Sd.Kfz. 135,

75 mm Pak 40/1 L/48 on Hotchkiss H 39 chassis,

75 mm Pak 40 L/46 (Sf) on Somua Schützenpanzerwagen,

75 mm Pak 40/1 L/48 on F.C.M. chassis.

Troop improvisations using the towed 75 mm Pak are not known, and probably not possible with the means at hand.

Upper right: 75 mm Pak 40 on the French Lorraine Tractor (Sd.Kfz. 135).
Right: In northern France, 1944. (1 x BA)

Left: A Marder I in Russia, winter 1942-43. It could carry 48 shells and was a useful vehicle in terms of engine and armament. (BA)

Below: Nopte the red "cat's-eye" on the muzzle protector. The fighting compartment was in the rear part of the vehicle; the crew consisted of five men. The gun had an elevation range from -5 to +22 degrees and a traverse of 33. Here the radio set is being tested. (BA)

Opposite page: There were 184 of them, also known by the designations GW 39 H and Marder I. Its height was 2.13 meters, width 1.88 and length 5.18 meters. (BA)

Above: The two upper photos show a Panzerjäger with the 75 mm Pak 40/1 on Hotchkiss H 39 chassis. The anti-aircraft machine gun is inside a protective cover. There were only 24 of these, and all saw service only in France. On the vehicle at left, a shear telescope can be seen. (BA)

Left: Only a few examples of this Panzerjäger are known. It has a 75 mm Pak 40 L/46 on the armored French Somua halftrack tractor—an unfortunate, very overweight vehicle.

On this page we see the combination of the 75 mm Pak 40 L/48 with the chassis of the French G.W F.C.M. There were only ten of these Panzerjäger built, and they all saw service only in France in 1944.

The right photo clearly shows the height and width of the vehicle. The fighting compartment was open at the top, and its crew consisted of five men. The armor plate was not very thick, and only safe from machine-gun and splinter fire. The rather underpowered engine made the vehicle ponderous. (1 x Sch.)

Marder II

The demand for more Panzerjäger that could match the Russian T-34 and KW family in mobility and penetrating power also led to the mounting of the German 75 mm Pak/Flak and 76.2 mm Pak (r) on the chassis of the Panzer II, which was somewhat outdated and not much used by that time.

These variations existed:

75 mm Pak 40/2 on Panzer II (A and F) chassis, Sd. Kfz. 131,

76.2 mm Pak (r) on Panzer II (D, E and F) chassis Sd.Kfz. 132.

The majority of the 76.2 mm Pak guns were fitted with a new inner barrel, which reduced the caliber to 75 mm, so that this gun could be used with German ammunition.

Upper and lower right: Mounting the 75 mm Pak 40/2 on the Panzer II, Type A and F chassis was a successful expedient. 531 of them were built. (1 x BA, 1 x Sch.)

This photo shows details of the Marder II very clearly. (Sch.)

The Marder II was used from 1942 on, on all front except Africa. The two photos at left show it in southern Russia in 1943. It proved itself splendidly. (BA)

Above: Here details of the barrel mounting can be seen.

All three photos were taken in Russia in 1943. The gun barrels show score rings, in white above, in black at right.

The Panzerjäger shown at right carries a paratroop crew and a camouflage of whitewash applied over old winter camouflage, a mixture often hard to find in the spring thaw. The extra pick and shovel on the back were not standard equipment. (BA)

These three photos are especially interesting. They show the first attempts to mount the 75 mm Pak on Panzer II C chassis. It is not known how many were made or whether they got into action. (BA)

This picture shows the excellent gun mounting on the Panzer II F chassis. Giving names or applying numbers, as in the right photo (IA were then the license-plate letters for Berlin) was first accepted in 1943—and never allowed in the German Army. (BA)

Both pages show the Marder II in winter action. All are painted white as winter camouflage. In the photo above, a German cross can be seen on the front (this was rare). The barrel brace and muzzle protector (left photo, opposite page) indicate an enemy-free area. The lower right picture shows the cloud of snow after a shot. (BA)

All three photos on this page show the Marder II with paratroop crews in Russia in 1943-44. The pictures at left show an additional MG 34 mount, which was used rarely. The same vehicle's camouflage (whitewash over winter paint) is interesting. (BA)

Opposite page: A weapon display in the northern sector of the eastern front in the summer of 1943 includes, in front, a Marder II with an interesting numbering and a very thin German cross. In the background is a French 75 mm Pak on the chassis of a Russian T-26 tank (see also page 41 of this book). (BA)

555

Opposite page: It could carry 37 shells, but the open fighting compartment could be very cold. The soldier on the front, of whom only the legs can be seen, is probably inspecting the steering. (BA)
This Marder II uses the chassis of Panzer II, Type D or E, recognizable by the four large road wheels and lack of return rollers. (BA)

Upper left: A factory-fresh Marder II on Gw II, Type D. Instead of the German 75, it used the Russian 76.2 mm Pak 36. This gun was sometimes fitted with a new inner barrel to reduce the caliber to 75 mm, so German ammunition could be used.

Above: Front view. It carried a four-man crew. Its shape was not as elegant as that of the other versions of Panzer II.

Left: A Marder II (76.2) on duty on the steppes before Stalingrad in 1942. On the horizon is the cloud of smoke from the burning city, visible for months.

Above: Note the bundles of switches on the track apron and on the bow.

Upper right: Well camouflaged. There were 316 of Types D and E made with the 76.2 mm gun. (BA)

Right: A Marder II (76.2) on a makeshift bridge at the Kuban bridgehead.

Other Panzerjäger of 75 to 125 mm Caliber

The German armaments industry developed—besides those already shown—other Panzerjäger, the construction of which (except for the "Nashorn") were generally halted for various reasons after testing of the prototypes. Several went on to troop testing, and some even reached the front lines. It is thus difficult to determine which types were only test models. They are shown here whether entire units were equipped with them or not.

Those that are known are:

75 mm Pak 40 L/48 on medium Schützenpanzerwagen, Sd.Kfz. 251/22
75 mm Pak 40 L/48 on heavy Armored Scout Car, Sd.Kfz. 234/4,
75 mm Pak 40/1 L/48 on Raupenschlepper Ost (RSO) chassis,
75 mm Pak 97/38 (f) on Russian T-26 chassis,
76.2 mm Pak 36 (r) on 5-ton medium towing tractor,
88 mm Flak 18 on 12-ton heavy towing tractor,
88 mm Flak 37 on special chassis,
88 mm Pak 43/1 on Gw III/IV,
88 mm Pak 43 on Ardelt special weapons carrier,
88 mm Pak 43 on Krupp-Steyr weapons carrier,
10 mm Kanone 18 on Gw III/IV, and
128 mm Pak 43 on Henschel VK 3001.

Above: A 75 mm Pak 40 L/48 on the medium Schützenpanzerwagen (Sd.Kfz. 251/22). Only a test model is known. Below: In 1944, several heavy armored scout cars (Sd.Kfz. 234) were equipped with the same gun. They did not reach the front.

The Panzerjäger shown on this page—75 mm Pak 40/1 on Raupenschlepper Ost (RSO) chassis—reached the anti-tank units of the grenadier divisions in 1944. It shows the typical makeshift construction: hasty work, no armor plate, small fighting compartment, no ammunition racks.

The photo at upper right shows that the side panels could be folded down.

Right: Weather protection—a tarpaulin over removable bows—was possible. The driver had to leave his seat during combat. (2 x Sch.)

This page shows a not very successful attempt to mount the Russian 76.2 mm Pak (without a muzzle brake) on the five-ton towing tractor and provide armor to protect the gun crew (but not the driver!).
On account of its weight, the armor plate was thin. The side entrance was very high and rather hard to use. Why this very high vehicle saw service in Africa (and nowhere else), where it was so conspicuous, is beyond understanding.
There were very few of these unfortunate vehicles.

Opposite page: Here is another somewhat adventurous attempt: The French 75 mm Pak 97/38 mounted on a Russian T-26 tank chassis. It can only have been built in a factory, as the troops lacked the means to do work like this—aside from the fact that there were no French guns to be captured in the east.

It was inevitable that the reliable 88 mm Flak gun be used as an anti-tank gun. The vehicle at upper left, with an 88 mm Flak 18 on a 12-ton towing tractor, was not successful.
Mounting the 88 mm Flak 37 on a special chassis (above) was technically better, but came to a stop for reasons cited in Volume 51 (page 44).

Left: A good development was the "Nashorn" Panzerjäger (at first called "Hornisse"). It had an 88 mm gun mounted on the chassis of Panzer IV with armor plate to protect the crew. Barely 500 of these were built.

The Ardelt firm in Eberswalde collaborated with the Rheinmetall-Borsig AG (which was responsible only for the weapons) to build a weapons carrier with the 88 mm Pak 43 L/71. The fighting weight of this gun carrier—which had also been planned to carry artillery guns—was 18 to 20 tons, depending on its armament.

The two photos above show a somewhat different version of the Ardelt weapons carrier with the 88 mm Pak 43 L/71.

Left: This prototype of a weapons carrier with the 88 mm Pak 43, built by the Krupp and Steyr firms, did not go into series production. Since it had a turning turret for all-around fire, one wonders why it was not regarded as a battle tank.

Both pictures show a 100 mm Kanone 18 (actual caliber 105 mm) mounted on the Panzer IV chassis. Only two of these were built, but, as the photos above and on the inside back cover show, they did see service. The soldiers nicknamed them "Fat Max".
The question remains of whether this is a Panzerjäger or armored artillery. Both uses were possible, but in practice it was used only as a tank destroyer.

The two photos on this page show "Stubborn Emil", with a 128 mm Pak mounted on the 3001 or 3601 test chassis built by Henschel. There were also only two examples of this model, and they too saw service in the east.

Only the Jagdtiger outweighed this heavy Panzerjäger.

Left: A great many score rings can be seen on the gun barrel.

Technical Data

Marder II (75 mm)

Engine	Maybach HL 62 TRM, carburetor
Cylinders	6
Bore x stroke	105 x 120 mm
Displacement	6191 cc
Power	140 HP at 2600 rpm
Torque	41.5 mkg
Compression	1:6.5
Carburetor	1 Solex 40 IFF II
Valves	Dropped, overhead cam driven by crown wheel
Main bearings	8, changeable cylinder liners
Cooling	Water, by pump
Battery	12 Volt 105 Ah
Generator	600 Watt
Power train	Rear engine, drive to tracks, two-plate dry clutch, central shift lever
Gearbox	ZF-Aphon SSG 46, 6 forward, 1 reverse, II to VI gears syncromesh
Body	Self-supporting armor plate, open fighting compartment, front and side armor
Tracks	2 tracks of 108 links (91 mm), front drive wheel, rear idler, 4 return rollers, 5 quarter-elliptic springs Steering & braking Clutch steering – mechanical steering brakes activated by two steering levers

General Data:

Track length	2400 mm
Track	1880 mm
Track width	300 mm
Dimensions with gun	4620 x 2270 x 2200 mm, length 6360 mm
Ground clearance	340 mm
Wading ability	920 mm
Turning circle	4.8 meters
Fighting weight	10,800 kg
Top speed	40 kph
Fuel consumption	110 liters on, 170 off road/100 km
Fuel capacity	170 liters (2 tanks)
Range	150 km on, 100 km off road
Crew	4 men
Armor Hull	14.5 to 35 mm, fighting compartment 8 to 10 mm
Armament	75 mm Pak 40/2 L/46

Nashorn

Engine	Maybach HL 120 TRM, carburetor
Cylinders	V-12 60-degree
Bore x stroke	105 x 115 mm
Displacement	11,867 cc
Maximum power	300 HP at 3000 rpm
Sustained power	265 HP at 2600 rpm
Torque	80 mkg at 2150 rpm
Compression	1:6.5
Carburetor	2 Solex 40 JFF II 2-stage downdraft off-road
Valves	Dropped, 1 camshaft per head, driven by crown wheels
Main bearings	7
Cooling	Water, by pump
Battery	Two 12 Volt 105 Ah
Generator	600 Watt
Starter	4 HP
Power train	Mid-engine driving tracks, three-plate dry clutch
Gearbox	ZF-Aphon SSG 77, 6 forward, 1 reverse, 2nd to 6th gears syncromesh
Transmission	Side shafts 4.00
Body	Self-supporting armored hull, armored fighting compartment with open top
Tracks	Two 103-link tracks, front drive wheel, rear idler, 8 small double road wheels in a row, 4 return rollers, 4 quarter-elliptic springs (one per pair of wheels)
Steering & braking	Mechanical planetary steering + mechanical steering brakes, two steering levers

General Data:

Track length	3800 mm
Track	2450 mm
Track width	400
Dimensions gun:	6200 x 2950 x 2940 mm, with 8440 x 2950 x 2940 mm
Armor plate	Hull front 30 mm, sides and rear 20 mm, fighting compartment 10 mm all around
Ground clearance	400 mm
Wading ability	800 mm
Turning circle	6 meters
Gross weight limit	24,000 kg
Load limit	3000 kg
Top speed	40 kph
Fuel consumption	240 liters on, 360 off road/100 km
Fuel capacity	600 liters
Range	260 km on, 160 km off road
Crew	4 to 5 men
Armament	88 mm Pak L/71 + 1 MG

Also from the Publisher